AF812146

Who Was Macho B

And What We Know About Jaguars

MARY JO NICKUM

Aquitaine Ltd
Phoenix, Arizona

Who Was Macho B

And What We Know About Jaguars

MARY JO NICKUM

Aquitaine Ltd
Phoenix, Arizona

CoCover design by JD Smith Designs

Copyright © 2016 Mary Jo Nickum

All rights reserved. This book or any portion thereof may not be reproduced or used in any manner whatsoever without the express written permission of the publisher except for the use of brief quotations in a book review.

Library of Congress Control Number: 2016953043

ISBN-13: 978-0-9980858-1-4
ISBN-10: 0-9980858-1-2

www.aquitaineltd.com

DEDICATION

This book is dedicated to young people everywhere who
don't like to or find it difficult to read and
who, therefore, live on the fringes of a happy, healthy life.
Learn to enjoy reading and the world can be yours.

Believe in yourselves

Why a Reluctant Reader Series?

Not all children read at the same level nor do all children enjoy reading. While a good many children do become excited and engaged in reading, especially in the primary grades, some are reluctant and disinterested. While a child may not show a natural interest in reading, this does not mean he/she cannot become a skilled and even enthusiastic reader. If the child has reached middle grade and is still disinterested, it's time to take action.

While any child, young or old, male or female can be a reluctant reader the largest number of unenthusiastic readers are adolescent boys. Research shows that a good number of boys who were avid readers in the elementary grades become disinterested in reading during their middle school years. However, there are a number of factors that may contribute to this shift—increasing complexity of material, peer pressure—one of the primary reasons seems to be they fail to see the connection between reading and "real" life.

School assignments, such as book reports, can become stressful for these students as well as parents and teachers. These children need material that is especially prepared to be relevant to the curriculum as well as to life. This material must be written to engage the reluctant reader, using images as well as text and presented in electronic and paper to fit classroom and leisure reading.

Books for the reluctant reader must be:

- Relevant to the curriculum (Common Core),
- Factual but engaging,
- Written to pique and hold interest,
- Presented in text and images,
- Available as ebook and in print, and
- Priced to be affordable for the individual student as well as bulk-priced to be attractive to schools.

Aquitaine, Ltd.'s Reluctant Reader Series, geared to middle grade readers, fills all these requirements. Our books have been thoroughly researched and edited by leading scientists and written by educators and librarians.

Illustration List

Table of Contents

Part 1

Who Was Macho B?

It all began at dusk one evening, under an outcropping of rock along the border between the US and Mexico, a male jaguar (*Panthera onca*) rose from his sleep, stretched, and silently but determinedly left the place he called home. A mysterious visitor from across the border, coming in the dark of night, a jaguar, dubbed by wildlife officials as Macho B, entered Cochise County, Arizona (Figure 1). Of course, he didn't realize he'd crossed a border, animals aren't aware of our human imposed boundaries. Upon arrival, Macho B found the territory available, except for Macho A, a young jaguar, also perhaps, looking for a new home.

Figure 1 Map of Arizona showing Cochise County

Macho A wasn't a real threat to Macho B. He would have to either "convince" him to leave or kill him. We don't know what Macho B did. Cats tend to first try to threaten an adversary with vocalizations and threatening postures before they resort to a physical encounter. Based on the lack of a camera record of Macho A after his initial sighting and the arrival and multiple sightings of Macho B, biologists concluded that Macho B either killed Macho A or ran him out of the territory.

Anyway, there was plenty of food—jackrabbits, deer and peccaries or javelinas, as they are also called, for Macho B. There was cover for him too, scrub, trees and rock caves everywhere, similar to the environment he'd left in Mexico (Figure 2).

Figure 2 The Sierra Madre Occidental of Northern Mexico

Macho B's first occurrence in Arizona was recorded by photograph in 1996 (Figure 3). He was not the first jaguar to take up residence in the United States, however. The occasional reports of jaguars between 1885 and 1965 indicated that a small but resident population did occur in southeastern Arizona. In December 1993, there was a report of a jaguar sighted on the Buenos Aires National Wildlife Refuge in the Baboquivari Mountains of southern Arizona. Could that have been Macho B as a young cat?

Figure 3 Jaguar on the border

There is no way of knowing exactly what Macho B did while he was in southern Arizona. We aren't even sure he spent the entire 13 years in Arizona. He may have traveled to southern New Mexico, Texas, or back to Mexico several times during those years. However, he seems to have spent some quality time in Arizona. He must have found enough food and shelter to sustain him. Like most other cats, jaguars are territorial. Macho B would likely have found suitable territory (Figure 4), marked it using urine, scat, and scratch marks, as do most felines, and was prepared to defend it from adversaries.

Figure 4 Chiricahua Mountains – suitable territory for jaguars

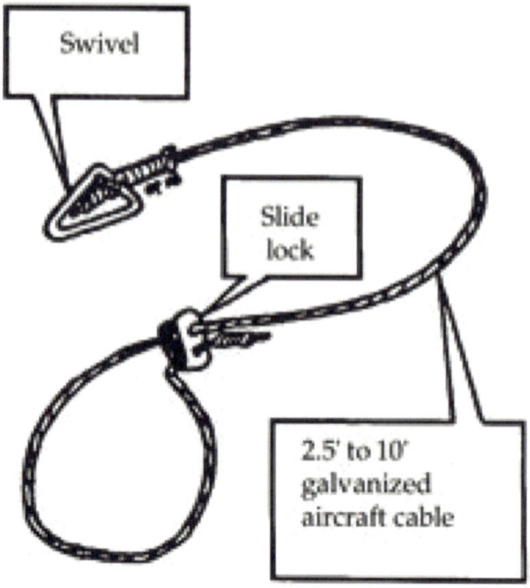

Figure 5a and 5b Leg-hold snare

On Feb. 18, 2009, a jaguar was captured, collared, and released south of Tucson. The jaguar, caught 'unintentionally' by the Arizona Game and Fish Department in a leg-hold snare (Figure 5a, b) set for mountain lions, was identified as Macho B. He was identified by the character of his spots. The spots on a jaguar are as individual as fingerprints of a human and Macho B had two easily recognized spots. These spots were rosettes (Figure 6), unlike the solid black spots normal to jaguars. The capture site was at approximately the 4000 ft. elevation, in a transition between desert grassland and oak woodland (Figure 7). Using a handling and capture method developed by leading jaguar experts, biologists took the opportunity to fit the cat with a lightweight satellite-tracking collar (Figure 8) in the hopes of learning more about the species' existence in Arizona. Immediately after the original capture, the cat appeared to be in fine health. He behaved exactly as biologists expected, he fled the capture site to a secluded area to recover (Figure 9). His every movement could now be tracked without interfering with his daily habits. Tracking-collar data transmitted in the days following his capture and collaring indicated that Macho B was doing well, traveling more than three miles after being released.

Figure 6 Macho B's signature rosettes

Figure 7 Forest-Oak transition

Figure 8 Radio tracking collar

Figure 9 Secluded area for recovery

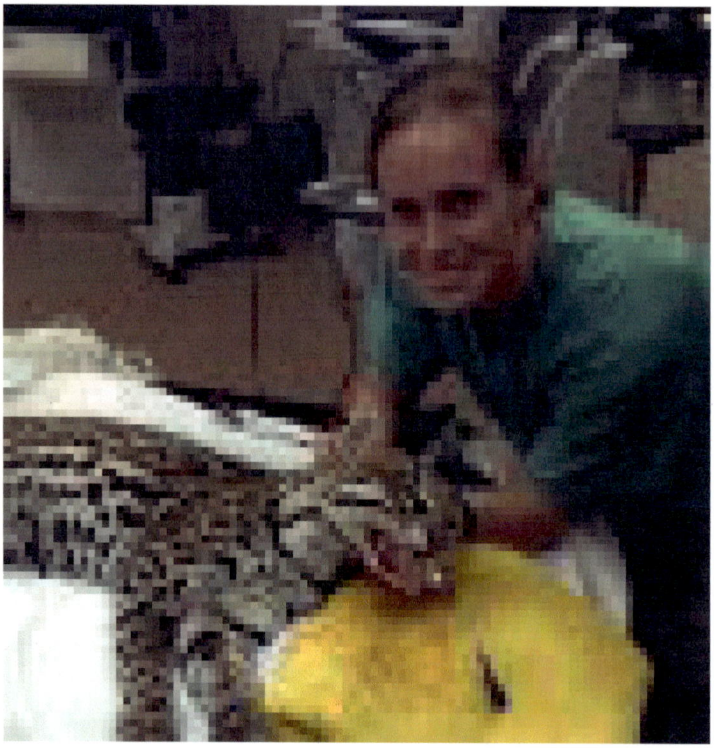

Figure 10 Kristen Phair, staff Veterinarian at the Phoenix Zoo (Not necessarily the one who treated Macho B)

However, in the following days it became clear that his movements were not normal— he was not moving. A field team, consisting of Arizona Game and Fish Department biologists and a wildlife veterinarian were sent to locate the animal and assess his overall condition. They determined the jaguar required recapture so his condition could be better evaluated by veterinarians at the Phoenix Zoo (Figure 10).

Through blood tests and a physical exam, zoo veterinarians confirmed the cat was suffering from severe and unrecoverable kidney failure. His condition of total kidney failure, though unfortunate, was not uncommon in old cats; he was determined to be about 15 years old. Acting on a recommendation from the zoo veterinarians, the U.S. Fish and Wildlife Service (USFWS) and the Arizona Game and Fish Department (AZGFD) determined that Macho B should be euthanized to end his "suffering."

Investigation into facts surrounding Macho B's demise has been concluded. Investigations of this nature are conducted by several government organizations. Concerned conservation groups and the interested citizens, who attended the hearings. The USFWS investigated because jaguars are an endangered species.

The death of Macho B was disappointing from every perspective. The wildlife conservation community looked forward to using the data acquired from him to learn more about how jaguars use borderlands (Figure 11) habitats and to improve borderlands conservation efforts. To the public at large, the erratic but magical presence of this wily recluse was a symbol of the incalculable value of the borderlands now and, hopefully, for future generations to come. Unfortunately, the mysteries that so many hoped would be revealed by Macho B will now remain secrets.

Figure 11 Borderland habitat

Nevertheless, we did learn some things from Macho B. He gave us his valuable DNA, a first for modern science, which will provide us genetic information about his origin, his relatedness to other jaguars and, thus, the viability of borderlands jaguars. Macho B was the oldest known wild jaguar in Arizona history and that is a clear testament to the **habitat** quality in southern Arizona (Figure 12). The fact that this jaguar was able to survive in this habitat for longer than any other jaguar in any other habitat, not only confirms jaguars can indeed thrive here, but also a huge network of public and private lands is currently being managed in a healthy and sustainable way.

Figure 12 Southern Arizona Chirichua jaguar habitat

By travelling across the border with Mexico, Macho B became an international ambassador and hero for jaguar conservation. He lived a long and magnificent life in a vast and majestic wilderness. His final days were tragic, though his contribution to our knowledge of jaguars was great.

A Jaguar Returns

In 2011, another adult male jaguar was seen near the proposed Rosemont Mine site (Figure 13a, b) in the mountains southeast of Tucson. This is the only jaguar known to live in the United States since the notorious jaguar, Macho B, died in Arizona in March 2009. "The continued presence of the jaguar in the Santa Ritas and elsewhere in the 'Sky Islands' mountain ranges of Southern Arizona (Figure 14) shows that jaguars belong in this region and underscores the need to protect their critical habitat," said Sergio Avila, a large cat biologist for the environmentalist Sky Island Alliance.

Figure 13a Rosemont Copper mine site

Figure 13b El Jefe

Figure 14 'Sky Islands'

According to the *Arizona Daily Star* newspaper, "Donnie Fenn, a professional guide based in Benson, Ariz., who specializes in mountain lion hunts—which are fairly common in Arizona—was taking his 10-year-old daughter out on her first lion hunt that morning when his pack of eight hounds took off in a frenzy. Before he knew it, he said, the dogs had a creature cornered in a tree, which he saw from afar with a telephoto lens was not the mountain lion he was looking for but instead an endangered jaguar."

"'I was scared,' Mr. Fenn said in a telephone interview. 'I didn't know if that thing was going to turn on me. I could feel its power. It was twice the size of a big mountain lion. It was definitely the experience of a lifetime.'" Mr. Fenn said his dogs were scratched pretty bad by the cornered jaguar, who probably had roamed north from Mexico.

In June, a helicopter pilot working along the border for the Federal Department of Homeland Security (Figure15) reported seeing a jaguar in the Santa Rita Mountains south of Tucson, officials said. Because the pilot had previously seen mountain lions, which are sometimes confused by the non-experts with jaguars, and was able to hover about 100 feet above the spotted cat and clearly describe it, wildlife experts took the report seriously. Yet biologists who went to the scene about a week after the spotting could not find tracks, hair or droppings from the animal, making it an unconfirmed sighting.

Figure 15 Border Patrol Helicopter

Mr. Fenn, made sure to confirm his run-in with the jaguar, which took place in an undisclosed mountain range in Cochise County. He crept up close after the jaguar was chased up a mesquite tree and took photos and a video of the animal (Figure 16). He also notified state wildlife officials, who were later able to find 15 hair samples left behind by the animal and a tree trunk that showed signs of being climbed by a large clawed animal. Experts believe Mr. Fenn saw an adult male jaguar that weighed about 200 pounds.

Figure 16 Jaguar treed

What Biologists Say

Biologists are 99.9 percent certain that Macho B came from Sonora, Mexico. In 1999, researchers found six live jaguars, including five females and one kitten, in the Sierra Madre Occidental in NE Sonora, Mexico (Figure 17; 130 miles south of the US/Mexico border). This is probably the source from which jaguars recorded in the U.S. in 1996 and 1997 originated.

Figure 17 Female Jaguar with kittens

In 1932, wildlife biologists thought the jaguar was a regular, but not abundant, resident in southeastern Arizona. By 1986, the jaguar was considered an uncommon resident species in Arizona. Perhaps, the jaguar ranged widely throughout a variety of habitats from Sonoran desert scrub upward through the subalpine conifer forest in Arizona (Figure 18).

Figure 18 Subalpine conifer forest

Though jaguars have been found in New Mexico and Texas, the most recent records of jaguars in the United States are from Arizona. In 1971, a jaguar was taken east of Nogales and, in 1986, one was taken from the Dos Cabezas Mountains (Figure 19). The latter individual reportedly had been in the area for about a year before it was killed.

Figure 19 Dos Cabezas Mountains

Questions of why he crossed the border and what he did while here cannot be answered with certainty. Large cats travel over great distances, sometimes in search of mates, prey, or defensible territory. Was Macho B searching for a new territory? Was he searching for more prey? Was his current situation in Mexico becoming "crowded"? It seems unlikely that six live jaguars could be considered a crowd, but Macho B might have found it to be so, especially if he had lost a battle for territory and a mate to another male jaguar. Finding new territory to the north seems likely because jaguars are found to the south all the way to Argentina (Figure 20); more territorial battles would have been necessary.

Figure 20 Jaguar-range-map

Jaguars hunt and consume a wide variety of prey. Peccaries (Figure 21) are considered the most attractive prey. Like most cats, jaguars usually catch and kill their prey by stalking or by ambush, biting through the nape of the neck. They are fond of turtles and tortoises and will flip a turtle on its back before scooping out the entire insides. Though they may break the shell, they are adept at consuming the whole turtle without breaking the shell. Jaguars hunt primarily at night and usually on the ground, though they can climb trees.

Figure 21 Collared peccary

Though jaguars are endangered and protected in Mexico, they can be serious predators of livestock. The northern most population of jaguars in the Americas inhabits the Sierra Madre Occidental in northeastern Sonora, in a landscape primarily used for grazing cattle (Figure 22). Illegal killing because of livestock predation is probably the main threat to this cat in this part of Mexico. Some Arizona ranchers were not happy to hear that a jaguar had moved into southern Arizona. They feared their cattle would be in danger because they thought if one jaguar came into the state, more would follow.

Figure 22 Grazing cattle *(Criollo)*

However, biologists who study these cats found this is not true. Healthy jaguars can range close to livestock without preying on them. Jaguar predation on livestock often occurs because of learned behavior of specific individual cats. Injuries to jaguars are the most common reason for this predator to turn to cattle. If the cat is injured or ill it may not have strength to hunt on rocky cliffs or crouch and spring to attack and kill fast moving prey. Lack of natural prey may cause jaguars to look for a calf as food. During periods of drought, ranchers often move cattle to grassy areas with dense surrounding vegetation (Figure 23), near a permanent water source. These same habitat conditions also attract wildlife, such as white-tailed deer, coatis, and peccaries, which increases the likelihood of the presence of predators. Jaguars prefer areas with dense vegetation because it provides stalking cover for ambush.

Figure 23 Grassland at forest edge

However, Macho B did not seem to attack cattle. At least, no rancher complained of calf losses during the time Macho B was in Arizona. In one case, a rancher agreed to work with the government and a conservation group, called the Jaguar Conservation Team (JagCT), to study and protect jaguars that come across the border. Apart from predation, ranchers had a more pressing concern when a jaguar was sighted in December 2001 in southern Arizona, approximately 5 miles north of the U.S./Mexico border. Ranchers in southern Arizona expressed concern the Federal government would begin to restrict land use practices. The JagCT held several community meetings with ranchers to inform them of jaguar habits and the restrictions in place to protect the jaguars and the ranchers.

What is Being Done?

The jaguar is listed under CITES (Convention on International Trade in Endangered Species) Appendix I. This means that all international trade in jaguars or their parts is prohibited. The International Union for the Conservation of Nature (IUCN) lists jaguars as Lower Risk, near threatened. The USFWS lists jaguars as 'Endangered'.

Major threats to the species are population fragmentation, deforestation and deliberate elimination. Overall, the trend in population is a decline. Any hunting of jaguars is prohibited in Argentina, Colombia, French Guiana, Honduras, Nicaragua, Panama, Paraguay, Suriname, United States, Uruguay and Venezuela, countries within the jaguar range. Hunting of jaguars is restricted to "problem animals" in Brazil, Costa Rica, Guatemala, Mexico and Peru. However, trophy hunting is permitted in Bolivia. The species has no legal protection in either Ecuador or Guyana.

Part 2

Were They Negligent with Macho B?

YES. The evidence is overwhelming that all agencies involved were willfully negligent in their treatment and study of Macho B. There are several reasons for this as pointed out by Janay Brun in her 2015 book, *Cloak and Jaguar; Following a Cat from Desert to Courtroom* (Figure 24). Ms. Brun's says the reasons can be boiled down to incompetence. All those involved either had a history of jaguar and cat mistreatment resulting in death of the cats or inability to make the best decision based on the available information. She provides harrowing details of capture by a leg snare rather than live trapping, which would have been less injurious for the animal.

Figure 24 Book Cover

Macho B had spent many hours attempting to climb the tree to which the snare line was attached (Figure 25), only to be jerked back to the ground when the length of the snare line was exhausted. Because this happened over a period of many hours, until the trap was finally checked, his leg and paw were scoured raw by the snare ring surrounding his leg (Figure 26) He was an old cat, perhaps 15-16 years old, and the stress of capture by leg-snare could have hastened the onset of kidney failure—a common problem for old felines, especially males.

Figure 25 Snared Jaguar

Figure 26 Snare wire

In *Cloak and Jaguar*, you'll meet the offending individuals. Ms. Brun worked with them on a daily basis. She was not, however, an employee of the Arizona Game and Fish Department or the US Fish and Wildlife Service. She worked for the conservation organization, Jaguar Conservation Team (JagCT) of the Borderlands Jaguar Detection Project (BJDP)—so did Emil McCain (Figure 27; Side Bar 1), the main character in the story of Macho B's demise.

"Aaah, the plot thickens" you say. Yes it does. Ms. Brun worked with McCain for years. She was new to the southwest, to jaguars and to the BJDT. She tried hard to learn from this seasoned researcher, respecting his experience and his overall good nature. As time went on and senseless events unfolded, Janay could take it no longer. She became a "whistleblower" (Side Bar 2), experiencing all the negative aspects of one person doing what's right in the face of strong resistance from state and Federal agencies, who were hard at work covering up the episode and its damning evidence.

Side Bar 1

Emil McCain pleads guilty

Well, I am saddened by this news but not surprised, as the wind was blowing that way. Emil shot himself in the foot professionally and he's going to pay the price. The entire Macho B saga has been one ugly scene and it probably isn't done yet. Here's the story from the AZ Star:

IN PLEA, MCCAIN ADMITS TELLING CO-WORKER TO ATTEMPT TO LURE MACHO B TO SNARE SITE

Tony Davis and Tim Steller Arizona Daily Star

The biologist at the center of the controversy over a jaguar's capture and subsequent death last year admitted Friday in federal court that he tried to snare the animal, known as Macho B.

Emil McCain pleaded guilty to a misdemeanor crime: illegal take of an endangered species. To "take" means to knowingly harass, harm, pursue, trap or kill, in violation of the Federal Endangered Species Act, court records show.

McCain, 31, of Patagonia, was immediately sentenced to five years' probation and fined $1,000.

In the plea agreement, McCain admitted an allegation that he had previously denied – that on Feb. 4, 2009, he told a female co-worker to place jaguar scat at snare sites in an effort to lure and capture the rare jaguar.

Previously, McCain had denied directing the woman, Janay Brun, with whom he worked on the nonprofit Borderlands Jaguar Detection Project, to place scat at the snare sites and said that if she did it, it was at her own initiative.

But Friday, he admitted to placing the scat or directing a woman to place the scat at three snare trap sites in an attempt to capture Macho B, the last known wild jaguar in the United States. He also admitted that he lacked permission to intentionally capture a jaguar.

The jaguar was found in the trap two weeks later by Arizona Game and Fish Department biologists working on what was supposed to have been a study of black bears and mountain lions.

"There was no authorization to intentionally capture a jaguar," McCain wrote in his signed plea agreement.

The aging jaguar was radio-collared by Game and Fish and released, but after it slowed down in the wild, it was recaptured 12 days later and euthanized due to kidney problems.

Under McCain's probation, he cannot be employed by or involved in any large-cat or large-carnivore project or study in the Unites States during that time.

When he leaves the country, he is allowed to be on unsupervised probation. McCain has worked as a biologist in Mexico, Spain and Costa Rica.

"He is done with the federal investigation and he has taken responsibility for whatever part he has in it," said his Tucson attorney, Alfred Donau III.

Encountered Friday afternoon at the federal courthouse in Tucson, McCain declined to talk about the case and walked away from a reporter. He had made his plea to Federal Magistrate Bernardo Velasco.

Until Friday, it was unclear if McCain would face charges. But in a procedure sometimes used by prosecutors, McCain was charged, pleaded guilty and was sentenced all in one day.

The federal government has had a criminal investigation since April 2009 into questions such as whether the Endangered Species Act was violated in the capture of Macho B. The criminal investigation is continuing, said Winn Hornbuckle, a U.S. Attorney's Office spokesman in Phoenix.

In late March 2009, Brun, a research technician with the Jaguar Detection project, alleged in an interview with the Arizona Daily Star that McCain had ordered her to put female scat at the trap site. McCain denied the charge at the time.

But Thornton "Thorry" Smith, then a biologist for Arizona Game and Fish, later told state investigators that McCain said scat had been placed at the sites of remote cameras near the eventual trap site. Originally, Game and Fish said the capture was accidental, occurring as part of its bear-lion study.

In his plea agreement Friday, McCain said he knew there had been recent evidence that a jaguar had appeared in the area of the snares. Photographs had been taken near the capture site in January 2009.

In his testimony to Game and Fish investigators, Smith admitted to lying to federal investigators about his role in the case – for which he was later fired – and concocting with McCain a coverup of the scat placement. But under the terms in which he testified, Smith can't be prosecuted for what he told the state.

Brun may now face prosecution. On Friday, she said she had been told by a U.S. Fish and Wildlife Service investigator that she would receive a summons today to appear in federal court next Thursday in connection with the jaguar case.

Nicholas Chavez, the Wildlife Service's Southwest law enforcement chief, said Brun is probably being summoned to face whatever charges the U.S. Attorney's Office plans to bring against her. U.S. Attorney's Office spokesman Hornbuckle declined to comment.

Brun said she was frustrated that McCain's sentence wasn't stiffer.

Because of Smith's statement about the coverup and because of McCain's involvement in other incidents in which jaguars died, "I don't think he should be allowed to be near any big cat or any animal in any country," said Brun.

In 2003, McCain worked on a jaguar capture in Sonora in which the animal died within a day after his release. McCain and another biologist acknowledged that they lacked proper training or equipment.

Last year, McCain worked in a project in the Mexican state of Yucatan in which two jaguars died shortly after their release from captivity. One died of causes never determined. The other was killed by people in the area, according to an account published by the International Union for Conservation of Nature.

Side Bar 2

What is a 'Whistleblower'?

A whistleblower is anyone who has and reports insider knowledge of illegal activities occurring in an organization. Whistleblowers can be employees, suppliers, contractors, clients or any individual who somehow becomes aware of illegal activities taking place in a business either through witnessing the behavior or being told about it. Whistleblowers are protected from retaliation under various programs created by the Occupational Safety and Health Administration (OSHA) and the Securities and Exchange Commission (SEC).

Figure 27 Emil McCain

In the wake of the whistleblowing episode, court proceedings followed. It was a long, drawn out affair, with the press ever anxious to expose the "truth" of the moment. In the end, though McCain and his cohorts lost jobs and were fined, no lasting penalties were incurred. Except for the whistleblower, which is the situation in more cases than not. Ms. Brun lost the will to continue in the southwest. She was, basically, burned out. That is the sad way this saga ended. The media has lost interest—moved on to the next "hot" issue; AZGFD continues to regard the incident as unimportant; the USFWS treats the matter as closed and does not accept any responsibility for it; and, the JagCT, a volunteer partnership between Arizona and New Mexico game and fish departments is keeping a low profile.

Not all biological field work is filled with as much excitement and drama as the story of Macho B but, having experienced the isolation and difficulties of a whistleblower, I can empathize with Ms. Brun and I admire her courage and honesty. The entire story of Macho B would make a fascinating and exhilarating movie—if anybody had the stamina and determination to make it. The movie would be a worthwhile contribution to "science for the public".

Part 3

What We Know About Jaguars

The jaguar is the largest felid in the New World (Figure 28). It is the only roaring cat native to the Americas. Body structure matches the general form of the cat (Felidae). Jaguars are muscular with shorter, massive limbs, a deep-chested body, and a relatively large, rounded head (Figure 29). The total length of adults varies greatly among subspecies, but generally ranges from 5 to 7 ft. for females and 5½ to 8 ft. for males, the longest recorded being almost 9 ft.; overall average is 6 ft. Females usually are 10 to 20 percent smaller than males. The jaguar is smaller than the lion (*Panthera leo*) or tiger (*P. tigris)* and slightly larger than the leopard (*P. pardus*). Individuals weighing 287 lb. to 348 lb. have been reported.

Figure 28 Largest felid in NA

Figure 29 Large, rounded head

The tail (Figure 30) is long, about one-third of the head plus the length of body. Shoulder height ranges from 2 ft. to 2½ ft. Ears are small, rounded, and black on the back with small white or buff central spots (Figure 31), as in the tiger. Hair is rather short and bristly.

Figure 30 Jaguar tail

Figure 31 Jaguar ears

The throat, chest, belly and interior parts of the feet have longer fur. **Condylobasal** length (distance from anterior border of the tooth socket of the median incisor to plane of posterior border behind the jaw hinge) of the adult skull may exceed 11 inches (Figure 32), but is usually between 7½ and 10 in.; it rarely is less than 7 in.

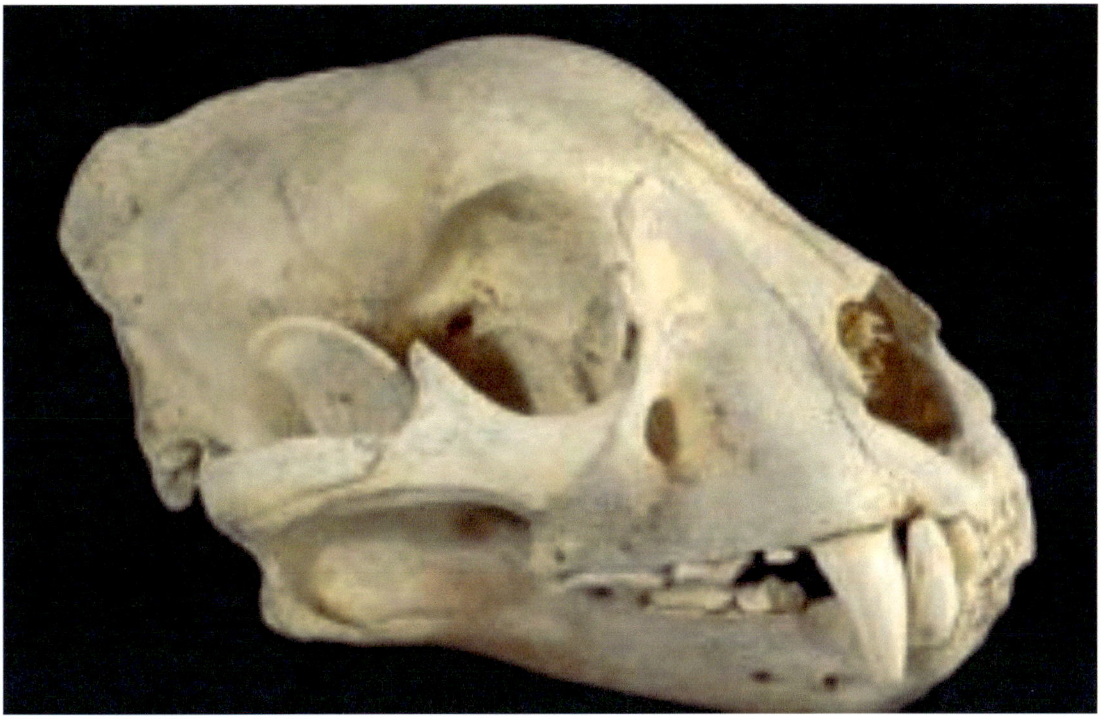

Figure 32 Jaguar skull

The skull is robust, relatively short, broad in the nose and mouth (more so in males than females), and wide in the cheekbones. The ridge of bone running lengthwise along the midline of the top of the skull may become well developed, especially in males and older individuals. The adult jaguar has a total of 26 permanent teeth (Side Bar 3).

Side Bar 3

Jaguar teeth arrangement

	Incisors	Canines	Premolars	Molars	Total U Total L	Total	Latin Names	Common Names
U	3-3	1-1	3-3	1-1	16	30	Felis	Jaguar, Mountain Lion, Ocelot
L	3-3	1-1	2-2	1-1	14			

U = upper jaw   L = lower jaw
1-1, 2-2, 3-3 = number of teeth type on right and left side of jaw

Overall, coloration varies from pale yellow to tan or reddish yellow. On any one individual, the color is nearly uniform over the mid-back area, although it pales to buff or white on the cheeks, side of neck, throat, and down the outer surfaces of legs, and lower parts of the flanks. Color is generally whitish on the throat, belly, and insides of limbs. Jaguars are spotted profusely at all ages; spots on the sides frequently form rosettes (Figure 33) that enclose one or several dots. This type of spotting is unique to jaguars. Along the middle of the back, the spots elongate and tend to merge. The spots often form more or less lengthwise rows on the flanks, roughly parallel to the stripe, running down the back. However, the spotting is extremely variable and often is not the same on right and left sides of an animal. The spots are solid on the head and neck and on the belly turn into large blotches. The tail is white underneath, with two or three black rings near the end and the tip is usually black.

Figure 33 Jaguar spots

Distribution

The home range of the jaguar, at one time, extended from the southwestern United States to southern Argentina (Figure 34). However, hunting pressure and habitat destruction have caused its disappearance from much of its former range. It is now officially gone from the United States, El Salvador, Uruguay and the developed areas of the Brazilian coast. It has a limited range in Mexico, Guatemala, and Argentina. It is not mentioned as presently occurring in Chile or Nicaragua and is endangered in Bolivia, Panama, Costa Rica and Honduras. It is considered rare in much of Peru, Colombia, and Venezuela.

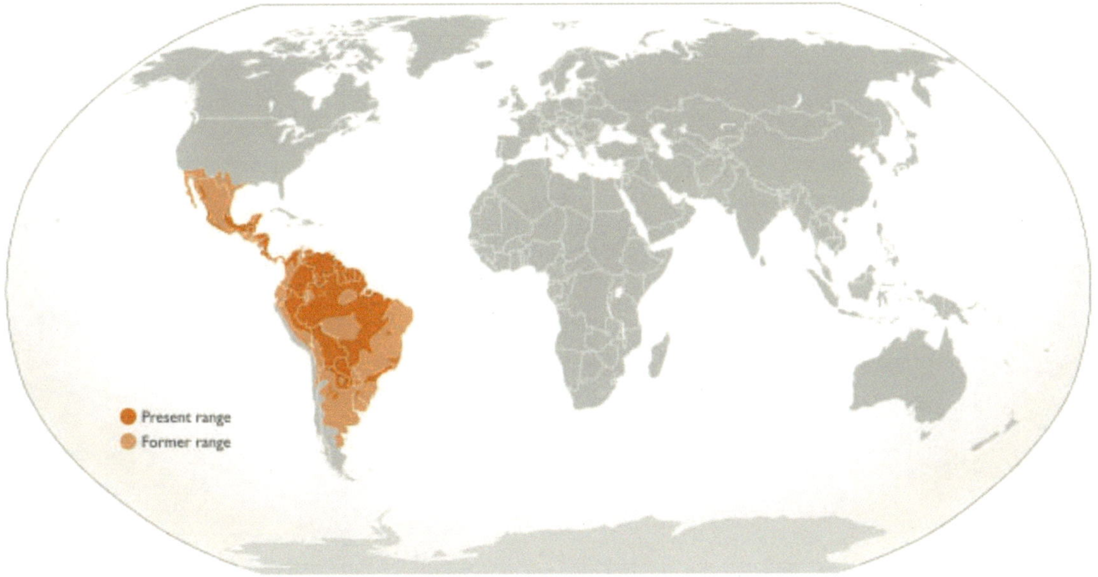

Figure 34 Range map

The largest remaining population now occurs in the Amazonian rainforest in Brazil, but jaguars are difficult to count because of their secretive nocturnal habits and dense forest cover. The jaguar is mainly tropical in distribution, usually selecting dense cover near water, although it has been found in many different habitats. They rarely are found at altitudes over 3,200 ft., however there are records of jaguars at 6,560 ft. in Venezuela, 6,889 ft. in northern Peru, and 8,858 ft. in Bolivia. Scant records exist to suggest its possible presence in eastern North America since the arrival of Europeans.

As you can see, from their original extensive geographic distribution, jaguars apparently are tolerant of a variety of environmental conditions. Most commonly, they are found in areas with considerable plant cover, a water supply, and sufficient prey. They have been reported from the rainforest, low-scrub jungle, lowland semi-deciduous forest, open tree and shrub woodland, swampy savanna, lagoons, marshland, floating islands of vegetation, thorn scrub and pampas/llanos, to deserts (Figure 35).

Figure 35 South American habitats

Behavior

Movements are similar to other members of the Panthera family, except jaguars sometimes appear more clumsy than the leopard. Their strides are about 19½ in. long, and often their tail is carried in an upwardly curved position as in the leopard. They climb trees quite well, but they do not do so as often as does the mountain lion. Jaguars like water and can swim exceedingly well, usually with their head and spine out of the water (Figure 36). They hunt and play in water, perhaps, even more frequently than the tiger. When emerging from the water, they shake the body and each of the paws separately. Many times, they have been seen swimming between islands and the shore. Postures, in general, are the same as in the other pantherines. They usually eat in a squatting or crouching position, frequently with paws on the kill (Figure 37). Their sleeping posture is often on the back (Figure 38), perhaps because of a relationship of body size to proportions.

Figure 36 Jaguar swimming

Figure 37 Jaguar eating kill

Figure 38 Jaguar sleeping on back

Jaguars usually catch and kill their prey by stalking or ambush and by biting through the nape of the neck (Figure 39), severing the spinal cord, as do most cats. For larger prey, such as cattle, they jump on the victim from the back or side, reach one paw over to grip the head and pull it around, so the animal unbalances and falls to the ground. The neck is usually broken in the fall; this bite does not break the neck. Sometimes they bite through the skull or, occasionally through the neck vertebrae of large prey. They rarely kill by biting the throat or by asphyxiation as do the tiger or leopard, respectively. The jaguar has developed a special bite for killing capybara (Figure 40). It bites from above or below, with its head at right angles to the victim and cracks the skull like a nutshell, sometimes inserting one upper and one lower canine into an ear of the victim. Most capybara taken by jaguars are young individuals, 4-years old or younger. Jaguars also are fond of turtles and will flip them onto their backs before proceeding to scoop out the entire insides. They are remarkably adept at consuming whole turtles *(Podocnemis)* or tortoises (*Geochelone*) in this fashion without breaking the shell (Figure 41). Jaguars also may break turtle shells to ease their consumption. To consume **caimans**, they turn them over and eat the unprotected underparts first; they may not eat the head, tail, and feet. Smaller prey species may be killed with one blow of the paw to the head.

Figure 39 Jaguar neck bite

Figure 40 Jaguar kills capybara

Figure 41 Jaguar consuming tortoise

Two eyewitness accounts as well as legends attest to the jaguar using its tail to attract fishes to the surface, whereupon it scoops them out of the water with a paw. Jaguars do capture fish in this manner; whether the tail is used consciously to attract fish is still unknown.

Jaguars sometimes move their kill to a more secluded or protected place, rarely in a tree; this may mean dragging large prey over one mile. Jaguars drag their prey while straddling it with their forelegs, with the carcass in their mouth, as does the tiger (Figure 42). They will cross rivers with their prey. Jaguars often consume the whole animal, even the skin, bones, hooves, and claws. They tend to consume the tongue, neck, and foreparts (brisket) of large animals first; often the rear end is not touched. They also like to eat the internal organs (heart, liver, spleen) but not the intestines, which they sometimes will move for some distance without breaking them. If disturbed and prey is abundant, they will abandon their prey without eating it. They may return to the kill a few hours later or as many as 3 days later. They usually make no attempt to hide their kill as do tigers and mountain lions.

Figure 42 Jaguar dragging kill

The jaguar is the least likely of the pantherines to attack man and is virtually undocumented as a man-eater. Only two cases were documented before 1914; other possible cases have been recorded but such occurrences are difficult to verify. There is no record of a jaguar systematically hunting man for food, as the tiger, lion, and leopard occasionally do. **Scavenging** is rare, but has been reported.

Although several jaguars may frequent the same area, they usually are solitary, except during mating or when the young are still dependent on their mothers. Young sibling pairs sometimes travel together. Jaguars use a land-tenure system similar to that used by tigers, leopards and mountain lions. Land tenure seems to be based on a "first come" system; when one animal dies, the ranges of the others shift. Jaguars primarily hunt at night and on the ground. Jaguars are not known to migrate regularly, although lone males have been known to roam hundreds of miles. Local adjustments of range may take advantage of seasonal changes in habitat. When pursued, they may travel 40 miles in one evening. Jaguars are known to use dens (Figure 43), especially when rearing young, although non-breeding adults also have been found to make dens in caves, canyons and ruins of ancient civilizations.

Figure 43 Jaguar den

Young captive jaguars display "threat and attack" behavior as well as "neck snapping"; these are more or less ritual forms of fighting, but probably form part of the learning process for the young (Figure 44). The first few months of vocalization consist solely of bleating; gurgling develops next, and then mewing, which replaces gurgling at 3 to 6 months. The pitch of calls lowers with age; 1-year old jaguars have gained the full repertoire, with the exception of some structured call sequences exclusively associated with reproductive behavior. **Prusten** is well developed in jaguars; it consists of an atonal, low-intensity, short (less than 1 sec.) snorting sound composed of several pulses, usually less than 10, divided by phases of decreasing intensity. These vocalizations are unique to jaguars, tigers, snow leopards (*Uncia uncia),* and clouded leopards *(Neofelis nebulosa*). By the end of the second year, jaguars develop a structured call sequence; these calls develop without learning from other jaguars. The most noted vocalization in the adult is the "grunt" sometimes described as a hoarse barking cough, consisting of five or six **guttural** notes. These vocalizations are done by both male and female and may serve for communication, especially during mating periods.

Figure 44 Jaguar kittens play-fighting

Scraping, urination, scent marking, deposition of feces, and tree raking may function in communication or as territory markers. In Belize, only scraping and feces deposition were recorded; these signs were noted particularly in areas of territory overlap. These scrapes averaged 14½ in. long and nearly 10 in. wide (Figure 45). Peruvian jaguars of both sexes go through periods of intense scraping. These signs were rarely noted in Brazil. Jaguars in Brazil select the morcequeira tree (*Andira inermis)* for tree raking, probably because of its thick trunk and rough bark. The bolletrie *(Manilkara bidentata)* is selected in Suriname. Tree scraping may only serve to remove loose claw sheaths as in the mountain lion and may not serve any communication function. Jaguars have been observed to spray urine backwards in the typical cat fashion. Urine spraying has been noted for the male jaguar but not for the female. Head-rubbing and claw-raking also have been noted as scent-marking behavior.

Figure 45 Jaguar claw scrapes

Mating behavior is the same as other members of the Felidae family. During mating, the female growls and the male licks the nape of her neck. When finished, the male bites her neck and jumps away while the female continues to roll around. Afterwards the male often drinks, urinates, or licks his genitals. During mating periods, fighting over a female is not common, although more than one male may follow a female. During **estrus**, there is a marked change in the female's behavior. Although her appetite does not decrease, she growls, rubs against objects, and rolls on the ground more frequently. The female will not let the male near her kittens because he may eat them, as tigers sometimes do. Up to weaning, the female will eat the feces of her kittens, which further prevents the kittens from danger of discovery by males.

Jaguars are sensitive to catnip (*Nepeta cataria*) even though the two are **allopatric** in the wild. Reproductive-aged adults are more sensitive than either aged or immature individuals.

Reproduction

The young are usually born with their eyes closed (Figure 46), opening after 3 to 13 days, averaging 8 days. However, it has been found that four of five kittens born in captivity opened their eyes on the first day (Figure 47). Young have a long, coarse, woolly coat, pale buff in color and heavily marked with round black spots, often with pale-colored centers. The kittens also have black stripes on their faces at birth. Kittens are about 15 in. long at birth and weigh between 1½ to almost 2 lb., averaging 1¾ lb. They gain weight at an average of 1½ ounces/day for the first 50 days, ranging from 1 to nearly 2½ ounces/day. The lower incisors erupt at 9 to 19 days, the upper incisors at 11 to 23 days, the upper canines around 30 days, and the lower canines at 36 to 37 days. The permanent teeth erupt in the same sequence as other Felidae, that is: upper teeth-P2, Ml, P4, P3; lower teeth-ml, p4, p3. The young first walk at about 18 days and may begin to take meat around 10 to 11 weeks, although suckling may continue for 5 to 6 months. They begin to follow their mother at approximately 6 weeks, although they may remain in the den for 2 months. They take on adult coloration around 7 months and remain with the mother for 1.5 to 2 years.

Figure 46 New-born Jaguar kitten with eyes closed

Sexual maturity ranges from 2 to 2.5 years to just over 3 years for females and 3 to 4 years for males. The age of last reproduction is about 8 years. Captives have lived to 22 years of age. In Belize, researchers found few wild jaguars over 11 years of age.

Figure 47 Jaguar kittens with eyes open

Relationship to Humans

The jaguar was worshipped by various South and Central American cultures, such as the Aztecs, Mayas, Olmecs, Toltecs, Zatopecs, and Nahualistas. There is a temple to the jaguar god in Tikal, Guatemala (Figure 48). The Olmecs reportedly deformed their heads to resemble this cat (Figure 49). In Paraguay, the celestial jaguar of the Guarani is believed to cause eclipses. The importance of the jaguar in the arts and myths of North American cultures probably has been underestimated.

Figure 48 El Gran Jaguar Temple

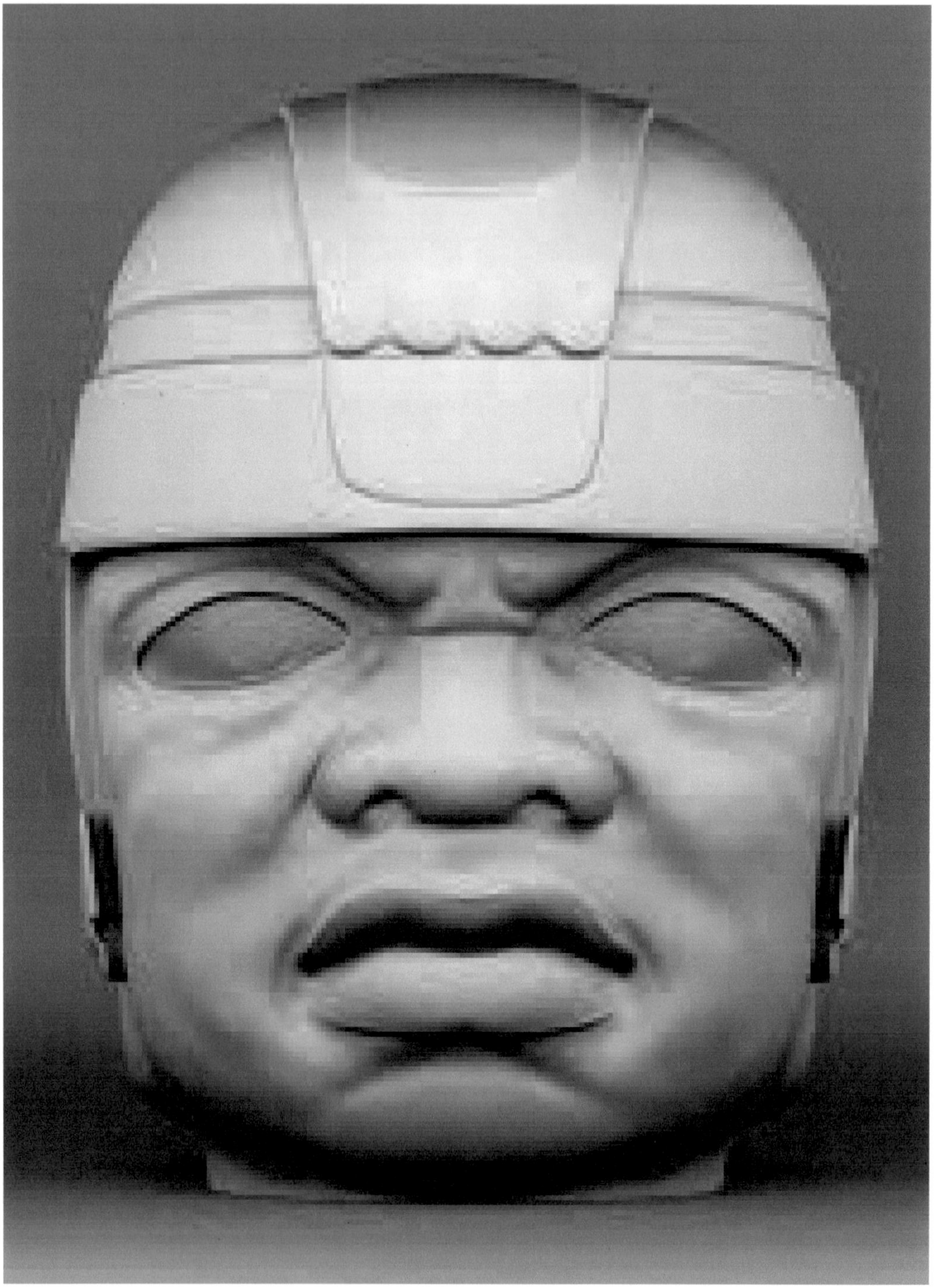

Figure 49 Olmec's deformed head

The name jaguar apparently is borrowed from one of the TupiGuarani languages; it was originally "yaguara", which means "wild beast that overcomes its prey at a bound". The Portuguese names for the jaguar are *ona verdadeira* and *ona pintada*. The Spanish name is *El tigre*. Native names for the jaguar include: yaguarete, yaui, zauat, uturuncu, nahuel, and acan gusu. Native names for **melanistic** (black) jaguars (Figure 50a, b) include: kuchi kudau, ming sha, and yaguaretehu.

Figure 50a Black jaguar with spots

Figure 50b Melanistic Jaguar

Unlike many other species, the jaguar faces no natural threat from rival cats or other predators—with the exception of humans. At the height of their decline in the 1960s and 1970s, more than 15,000 jaguars were killed each year for their beautiful fur. While trade in jaguar fur has been reduced by consumer awareness campaigns, they continue to face pressure from hunters. In addition, the loss of habitat is increasing the frequency of contact between people and jaguars. There are about 15,000 jaguars alive in the wild today. If jaguars are to exist in the future, scientists believe it will be through a combination of special protected areas and increased community awareness.

One more important point, jaguars are the least likely of the large cats to attack man. In fact, there is virtually no evidence of man-eating among jaguars, in contrast to data for lions, tigers and leopards.

Ecology

As implied by their original, extensive geographic distribution, jaguars apparently are tolerant of a variety of environmental conditions. They most commonly are found in areas with considerable plant cover, a water supply and sufficient prey. They have been reported from the following habitat types: rainforest, low scrub jungle, lowland semi-deciduous forest, open tree and shrub woodland, swampy savanna, lagoons, marshland, floating islands of vegetation, thorn scrub, pampas/llanos, and deserts. In Brazil, jaguars utilize all types of vegetation.

Mountain lions and jaguars are known to have overlapping ranges and have little interaction; indeed, they seem to mutually avoid one another. This is not the case for jaguar and ocelot, however. They have been photographed crossing the Mexico-Arizona border on numerous occasions within short periods of each other (Figure 51). When the ranges of mountain lions and jaguars overlap, mountain lions are more abundant in drier areas and jaguars select wetter areas.

Figure 51 Jaguar crossing border

Side Bar 4

Did you know?

- One of the jaguar's most distinctive features is its repertoire of calls, including a variety of snarls, growls, and "roars," which are really a series of hoarse coughs or grunts.
- The eyes of jaguars have round pupils and irises that range in color from golden to reddish yellow. Very young kittens have blue eyes, as with all felines.
- Jaguars are one of the few species of wild cats that have melanistic individuals, yet black jaguars have never been seen in their northern range.
- Jaguars are rapid runners but tire quickly and rely on proximity rather than sustained speed while hunting.
- Jaguar tracks (Figure 54) are unmistakable in their roundness, both in the pad and in the four toes that touch the ground.
- Jaguars have huge eyes (Figure 55), the largest of all carnivores relative to head size.
- Jaguars were once presumed to be nocturnal, but recent studies show they are also active during the day, with peaks of activity at dawn and dusk.

https://www.northernjaguarproject.org/who-we-are/

Jaguar Conservation

In Arizona and New Mexico, a state-led Jaguar Conservation Team (JagCT) is working to protect and conserve a species that many people do not even know is native to the United States. Created in 1997, JagCT is a voluntary partnership among state, federal, and local government agencies, private individuals, and other entities with an interest in jaguar conservation. Their efforts and those of colleagues in Mexico are helping create a more promising future for the jaguar in the U.S.-Mexico borderlands.

In 1996, independent discovery of two different jaguars changed prevailing perceptions about the species' presence in the United States. In March 7, 1996, houndsman/rancher Warner Glenn discovered and photographed an adult male jaguar in southwestern New Mexico (see *Eyes of Fire*; Figure 52). It was the first jaguar documented in the United States since two were killed illegally in Arizona in 1971 (near Nogales) and 1986 (Dos Cabezas Mountains). Then, amazingly, on August 31, 1996, yet another houndsman, Jack Childs, discovered a different adult male jaguar in south-central Arizona (see *Ambushed on the Jaguar Trail: hidden cameras on the Mexican Border)*. In no small way, Warner Glenn's and Jack Childs' discoveries and their passion for jaguars are responsible for inspiring the borderlands jaguar conservation effort in which they continue to participate today.

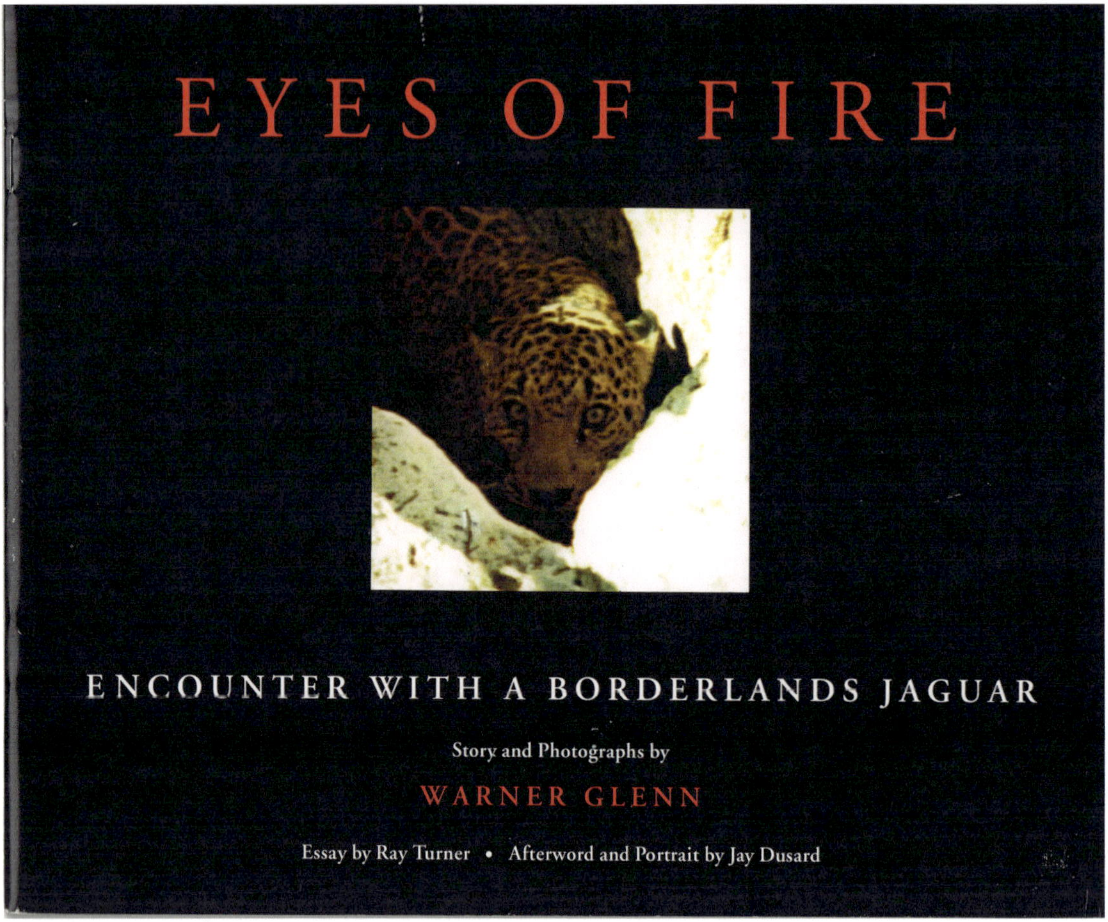

Figure 52 *Eyes of Fire* **Book cover**

By 2009, JagCT monitoring efforts, including continued vigilance by Glenn and Childs, had confirmed occurrence of four different adult male jaguars (possibly as many as six) since 1996 in the borderlands of southern Arizona and southwestern New Mexico. No females or sub-adult males were documented during that period. The last documented female jaguar occurrence in Arizona was in 1963, when one was shot near Big Lake, White Mountains. Other females have been reported for Arizona, notably one with a kitten in 1910 and another with two kittens in 1906. No records exist for occurrence of any female jaguars in New Mexico.

The monitoring data are not sufficient to reveal if even one of the jaguars documented between 1996 and March 2009 was continuously present in the United States, even within a single year. However, the data are sufficient to confirm that the jaguar Jack Childs observed on August 31, 1996 was present in Arizona many times between that original sighting and its death on March 2, 2009. Macho B was photographed by remote camera-traps (Figure 53) dozens of times over that period, in a home range of about 500 square miles—just considering the Arizona component. Macho B also was documented crossing the border in both directions (Figure 54). Nobody knows how widely he traveled in Mexico, before returning to Arizona.

Figure 53 Macho B crossing the border

Although there are different challenges in different areas, loss and fragmentation of habitat and illegal killing continue to threaten jaguars throughout much of their range. The USFWS listed jaguars outside the United States as an endangered species in 1972. The species was protected under the Convention on International Trade in Endangered Species of Wild Fauna and Flora (CITES) in 1973. In 1997, with enough solid biological evidence to indicate the Arizona and New Mexico borderlands are a legitimate (albeit tiny) part of the jaguar's current range, USFWS extended the federal listing to include the United States. Jaguars are also protected by state law in Arizona and New Mexico.

Figure 54 Jaguar track

Jaguars are known to be far ranging. Movements of 500 miles have been recorded. If food is abundant, they may become sedentary and range over only a few square miles. Like most cats, this species is territorial and marks its boundaries with scents. Jaguars roar to announce their presence to other jaguars; no other cat native to North America roars.

For jaguars to thrive or even to persist in Arizona, a few modest needs must be met. They must be protected from being killed. They must have an adequate prey base. And they must have movement corridors to connect with source populations in northern Mexico. Abundance of available prey, and suitable resting sites, are more important than any particular vegetation type to this wide-ranging species. The core population of jaguars in northern Mexico must also be sufficiently large to provide for dispersal into the U.S.-Mexico borderlands. Field research, especially on habitat use and movement patterns, in Arizona, New Mexico, and Mexico is underway to provide a sound scientific basis for management decisions, but much more work is needed.

The AZ-NM/Mexico jaguar conservation efforts include owners of private lands, ranchers on public lands, nongovernmental organizations, scientists, and state and federal agencies in Arizona, New Mexico, and Mexico. They are all working to identify and meet jaguar conservation needs throughout the borderlands. Progress has been made since 1997. However, a variety of major conservation needs still exist:

1. Jaguars in the borderlands must be protected from unlawful killing,
2. State criminal penalties in Arizona and both civil and criminal penalties in New Mexico should be commensurate with federal penalties,
3. Borderlands jaguars must be studied so that all conservation decisions about them are better informed by credible data about their occurrence, diet, behavior, and habitat use,
4. Borderlands habitats must be managed in ways that provide the basic elements that jaguars must have, including native prey, cover/shelter, water and natural corridors that enable movement back and forth across political boundaries and from one area to another within those boundaries,
5. Jaguar conservation efforts in the United States must continue to integrate more effectively with those in Mexico and,
6. Outreach is needed to inform and educate agencies and the public about borderlands jaguars and their legal protections and ecological needs.

Hopefully, the conservation efforts being designed and implemented today by JagCT and its cooperators, including the public and Mexico, will help provide future generations in the United States and Mexico with a unique gift—continued existence of jaguars that roam freely across the U.S.-Mexico border and throughout the borderlands, including California and Texas.

Side Bar 5

PROJECT OVERVIEW

The Borderlands Jaguar Detection Project is a non-invasive study designed to detect jaguars in the remote mountains along the border between southern Arizona and Sonora, Mexico. Cameras have been placed along the most probable travel routes in several small mountain ranges and riparian areas along the U.S./ Mexico border. These cameras detect heat in motion and photograph any animals that pass by. They record dates and times which help researchers understand the behavior and ecology of the animals.

The information gained from this study will help game managers and biologists make sound management decisions concerning conservation of the jaguar and its habitat. This project identifies core areas of suitable habitat and travel corridors that must be protected for the jaguar. This information will prove instrumental in jaguar conservation in the northern portion of its range and will help allow the jaguar to return to its home range in the American Southwest.

Side Bar 6

Panthera onca

http://dx.doi.org/10.2305/IUCN.UK.2008.RLTS.T15953A5327466.en

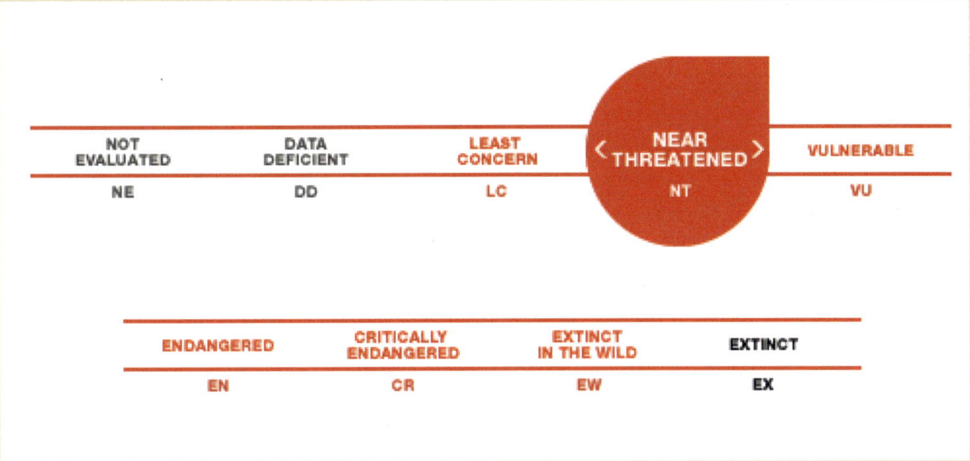

- Summary
- Classification Schemes
- Images & External Links
- Bibliography
- Full Account

Taxonomy
Assessment Information
Geographic Range
Population
Habitat and Ecology
Threats
Conservation Actions

Taxonomy

Kingdom: Animalia

Phylum: Chordata

Class: Mammalia

Order: Carnivora

Family: Felidae

Scientific Name: Panthera onca

Species Authority: (Linnaeus, 1758)

Common Name(s):

English – Jaguar

Spanish – Tigre Real, Tigre Americano, Otorongo, Yaguar, Yaguarete

Synonym(s): Felis onca Linnaeus, 1758

Taxonomic Notes: Although eight subspecies have been recognized (Seymour 1989), genetic (Eizirik et al. 2001, Ruiz-Garcia et al. 2006) and morphological (Larson 1997) analysis finds no support for the existence of discrete subspecies. Rather, variation is clinal - the greatest differences are found between Jaguars at the latitudinal extremes of their range. While not elevating the regional differences to the subspecies level, Eizirk et al. (2001) found evidence for four incompletely isolated phylogeographic groups: Mexico + Guatemala, southern Central America, northern South America, and South America south of the Amazon river. Similarly, Ruiz-Garcia et al. (2006) found that the Andes mountains incompletely isolates Jaguar populations in Colombia.

Assessment Information

Red List Category & Criteria: Near Threatened ver 3.1

Year Published: 2008

Date Assessed: 2008-06-30

Assessor(s): Caso, A., Lopez-Gonzalez, C., Payan, E., Eizirik, E., de Oliveira, T., Leite-Pitman, R., Kelly, M. & Valderrama, C.

Reviewer(s): Nowell, K., Breitenmoser-Wursten, C., Breitenmoser, U. (Cat Red List Authority) & Schipper, J. (Global Mammal Assessment Team)

Justification: Jaguar extent of occurrence was estimated at 8.75 million km^2, and probability of long-term survival was estimated as high for over 6 million km^2 (Sanderson *et al.*, 2002). Much of the jaguar's remaining habitat is the rainforest of the Amazon basin, which is of relatively low suitability (Torres *et al.* 2007). The jaguar is still an abundant species, but is threatened by habitat loss and persecution. Due to loss of habitat, poaching of prey and fragmentation of populations across portions of the range, this species is considered to be Near Threatened. If threats continue at the current rate the species will likely qualify for VU A2cd or A3cd in the near future.

Previously published Red List assessments:

2002 – Near Threatened (NT)

1996 – Lower Risk/near threatened (LR/nt)

1990 – Vulnerable (V)

1988 – Vulnerable (V)

1986 – Vulnerable (V)

1982 – Vulnerable (V)

Geographic Range

Range Description:

The jaguar is the largest cat of the Americas, and the only living representative of the genus *Panthera* found in the New World (Nowell and Jackson 1996). Historically it ranged from the southwestern US (where there are still some vagrants close to the Mexican border) through the Amazon basin to the Rio Negro in Argentina. Its extent of occurrence (EOO) is estimated at 8.75 million km^2, with its stronghold the rainforest of the Amazon basin, which comprises 88% of its EOO. However, ecological models indicate that much of the Amazon is of low suitability for the jaguar, compared with the Pantanal, Paraguayan Chaco, and Caatinga (Torres *et al.* 2007). The Jaguar has been virtually eliminated from much of the drier northern parts of its range, as well as northern Brazil, the pampas scrub grasslands of Argentina and throughout Uruguay. It is now estimated to occupy only about 46% of its historic range (Sanderson *et al.* 2002). Populations in Colombia are divided by the Andes (Ruiz-Garcia *et al.* 2006).

Sanderson *et al.* (2002) presented a group exercise to define the most important areas for conservation of viable jaguar populations (Jaguar Conservation Units or JCUs). These 51 areas add up to 1.29 million km^2, or 13% of jaguar range.

Countries occurrence:

Native: Argentina; Belize; Bolivia, Plurinational States of; Brazil; Colombia; Costa Rica; Ecuador; French Guiana; Guatemala; Guyana; Honduras; Mexico; Nicaragua; Panama; Paraguay; Peru; Suriname; United States; Venezuela, Bolivarian Republic of

Regionally extinct: El Salvador; Uruguay

Upper elevation limit (metres): 3000

Range Map: Click here to open the map viewer and explore range.

Population

Population:

Sanderson *et al.* (2002) found that 70% of estimated jaguar range (over 6 million km^2) was considered to have a high probability for survival. Most of this area consists of the Amazon basin rainforest, and adjoining areas of the Pantanal and Gran Chaco. However, while the latter two were found by ecological models to be highly suitable for jaguars, the Amazon was found to be of low suitability (Torres *et al.* 2007). Jaguar densities in the Brazilian Pantanal are estimated at 6.6-6.7 adults per 100 km^2 (Soisalo and Cavalcanti 2006), and in the Bolivian Gran Chaco 2.2–5 per 100 km^2 (Maffei *et al.* 2004). In the Amazon basin in Colombia, jaguar density was estimated at 4.5/100 km^2 in Amacayacu National Park and 2.5/100 km^2 in unprotected areas (Payan 2008). In Madidi National Park in the Bolivian Amazon, density was estimated at 2.8/100 km^2 (Silver *et al.* 2004).

Other high probability areas for long-term jaguar persistence include tropical moist lowland forest in Central America: the Selva Maya of Guatemla, Mexico and Belize; and a narrow strip of the Choco-Darien of Panama and Colombia to northern Honduras. Densities in the Belizean Selva Maya rainforest were estimated at 7.5-8.8/100 km^2 (Silver *et al.* 2004). The Talamanca Mountains of Costa Rica and Panama also host a populations, but the long term persistence is uncertain (Gonzalez-Maya *et al.* 2007).

Eighteen percent of jaguar range (1.6 million km^2) was estimated to have medium probability of long-term survival. These areas are generally adjacent to high-probability areas and include a large portion of the northern Cerrado, most of the Venezuelan and Colombian llanos, and the northern part of Colombia on the Caribbean coast. In Central America and Mexico, medium-probability areas include the highlands of Costa Rica and Panama, southern Mexico, and the two eastern mountain ranges of Mexico, Sierra de Taumalipas and the Sierra Madre Oriental.

The remainder of jaguar range was classified as low probability for jaguar survival, and of most urgent conservation concern. These areas include the Atlantic Tropical Forest and Cerrado of Brazil; parts of the Chaco in northern Argentina;

the Gran Sabana of northern Brazil, Venezuela and Guyana; parts of the coastal dry forest in Venezuela; and the remainder of the range in Central America and Mexico.

Some of the most important areas for jaguar conservation (Jaguar Conservation Units) fell within parts of jaguar range where probability for long-term survival was considered low, and so represent the most endangered jaguar populations. These include the Atlantic Forests of Brazil, northern Argentina, central Honduras, and the Osa peninsula of Costa Rica (Sanderson *et al.* 2002). The Atlantic Forest subpopulation in Brazil has been estimated at 200+/- 80 adults (Leite *et al.* 2002). Jaguar populations in the Chaco region of northern Argentina and Brazil, and the Brazilian Caatinga, are low-density and highly threatened by livestock ranching and persecution (Altrichter *et al.* 2006, T. de Oliveira pers. comm. 2008).

Current Population Trend: Decreasing

Additional data: *Population severely fragmented:*

Habitat and Ecology

Habitat and Ecology: The species is strongly associated with the presence of water. Habitats range from rainforest to seasonally flooded swamp areas, pampas grassland, thorn scrub woodland, and dry deciduous forest (Nowell and Jackson 1996). In Belize, jaguars are reportedly more abundant in lowland areas of relatively dense forest cover with permanent water sources than in open, seasonally dry forests. Although jaguars have been reported from elevations as high as 3,000 m (Brown and Lopez Gonzalez 2001), they typically avoid montane forest, and have not been found in the high plateau of central Mexico or above 2,700 m in the Andes. Jaguars take a wide variety of prey species but large-sized ungulates are preferred when available (Nowell and Jackson 1996).

A 13 year old wild female was found with cub (Brown and Lopez-Gonzalez 2001).

Density estimates ranged from 1.7-4 adults per 100 km^2 in studies in Brazil, Peru, Colombia and Mexico summarized by Sunquist and Sunqujist (2002). Density estimates by Silver *et al.* (2004) from five different study sites ranged from 2.4-8.8 adults per 100 km^2, with the highest densitiy found in Belize's Cockscomb Basin Wildlife Reserve (rainforest), a density similar to the 6-8 per 100 km^2 found by Rabinowitz and Nottingham (1986). That study found home ranges of females of 10 km^2, overlaped by male home ranges which varied from 28-40 km^2 and also overlapped extensively. In other areas jaguar home ranges have been over 1,000 km^2 (T. de Oliveira pers. comm. 2008).

Soisalo and Cavalcanti (2006) used GPS-telemetry to check density estimates derived from a common camera trap methodology in the Brazilian Pantanal, and cautioned that the method may over-estimate population size. Telemetry data indicated a density of 6.6-6.7 adult jaguars per 100 km^2, while densities derived from Maximum Distance Moved (MMDM) extrapolations from camera trap captures were higher at 10.3-11.7/100 km^2.

Jaguar densities in the Paraguayan Gran Chaco are 2.27–5.37 per 100 km^2 (Cullen Jr. *et al.* in submission), and in the Colombian Amazon, 4.5/100 km^2 in Amacayacu National Park and 2.5/100 km^2 in unprotected areas (Payan 2008). In Brazil, densities are 2 per 100 km^2 in the savannas of the Cerrado, 3.5/100 km^2 in the semiarid scrub of the Caatinga, and 2.2/100 km^2 in the Atlantic Forest (Silveira 2004, in litt. To T. de Oliveira 2008).

Systems: Terrestrial

Threats

Major Threat(s): Deforestation rates are high in Latin America and fragmentation of forest habitat isolates jaguar populations so that they are more vulnerable to human persecution (Nowell and Jackson 1996). People compete with jaguars for prey, and jaguars are frequently shot on sight, despite protective legislation (Nowell and Jackson 1996). An estimated 27% of jaguar range has a depleted wild prey base (WCS 2008). Jaguars are also known to kill cattle, and are killed by ranchers as pest species. The vulnerability of the jaguar to persecution is demonstrated by its disappearance by the mid-1900's from the south-western US.

Commercial hunting and trapping of jaguars for their pelts has declined drastically since the mid-1970's, when anti-fur campaigns and CITES controls progressively shut down international markets (Nowell and Jackson 1996). However, although hunting has decreased there is still demand for jaguar paws, teeth and other products.

Conservation Actions

Conservation Actions: Included on CITES Appendix I. The jaguar is fully protected at the national level across most of its range, with hunting prohibited in Argentina, Brazil, Colombia, French Guiana, Honduras, Nicaragua, Panama, Paraguay, Suriname, United States, and Venezuela, and hunting restrictions in place in Brazil, Costa Rica, Guatemala, Mexico and Peru (Nowell and Jackson 1996).

With habitat fragmentation a major threat, and taxonomic research suggesting little significant differences among jaguar populations, an ambitious program has been launched to conserve a continuous north to south habitat corridor through the species range (Rabinowitz 2007).

Addressing livestock management and problem animal issues is a high priority for conservation effort in many jaguar range countries.

Citation: Caso, A., Lopez-Gonzalez, C., Payan, E., Eizirik, E., de Oliveira, T., Leite-Pitman, R., Kelly, M. & Valderrama, C. 2008. *Panthera onca*. The IUCN Red List of Threatened Species 2008: e.T15953A5327466. http://dx.doi.org/10.2305/IUCN.UK.2008.RLTS.T15953A5327466.en. Downloaded on **24 April 2016**.

Glossary

Allopatric occupying separate, non-overlapping geographic areas.

Caimans a member of order Crocodylia (or Crocodilia), which are amphibious carnivores. They live along the edges of rivers and other bodies of water

Capybara largest rodent in the world.

Condylobasal distance from front of the tooth socket of the middle incisor to plane of back border behind the jaw hinge

Estrus a regularly recurrent state of sexual excitability during which the female of most mammals will accept the male and is capable of conceiving.

Guttural having a harsh grating quality, as certain sounds produced in the back of the mouth

Habitat the natural environment in which a species or group of species lives.

Melanistic an increased amount of black or nearly black pigmentation (as of skin, feathers, or hair) of an organism

Prusten a vocalization made by members of the Panthera family, often referred to as chuffing. It is a non-threatening vocalization often used between two cats who are greeting each other, during courting, or simply by a mother comforting her kittens.

Scavenging to feed on dead or decaying matter, especially of animals.

Taxonomy the process or system of describing the way in which different living things are related by putting them in groups

Resources

Altrichter, M., G. Boaglia and P. Perovic. 2006. The decline of jaguars *Panthera onca* in the Argentine chaco. Biology Digest 40(3): 302-309.

Attenborough, D. 2002. The Life of Mammals. Princeton University Press, Princeton, NJ.

Azevedo, F. 2008. Food habits and livestock depredation of sympatric jaguars and pumas in the Iguaçu national park area, south Brazil. Biotropica 40/4: 494-500.

Baker, W., S. Deem, A. Hunt, L. Munson and S. Johnson. 2002. Jaguar species survival plan. Pages 9-13 *In*: C. Law, editor. Guidelines for Captive Management of Jaguars, Vol. 1(1). Jaguar Species Survival Plan Management Group, Fort Worth, Texas.

Brun, Janay. 2015. Cloak and jaguar. Relham, LLC, Silver City, NM

Carrillo, E. 2007. Tracking the elusive jaguar. Natural History 116(4): 30-34.

Christie, Bob. 2009. Environm,antal group sues Ariz. To bar trapping od jaguars. http://www.abqjournal.com/news/state/apazjaguarlawsuit09-24-09.htm

Crawshaw, P. and H. Quigley. 1991. Jaguar spacing, activity and habitat use in a seasonally flooded environment in brazil. Biological Sciences 223/3: 357-370.

Davis, Tony. 2011. "G&F woes preceded jaguar capture, personality clashes, communication, policy lapses plagued agency." Arizona Daily Star, February 6 http://tucson.com/news/science/environment/g-f-woes-preceded-jaguar-capture/article_d594fa7b-81e4-5f36-8bb7-027ce11d83ec.html

Davis, Tony. 2011. "Researcher's status questioned in jaguar case, testimony indicates he was working for state; game officials deny it." Arizona Daily Star, January 30 http://tucson.

com/news/local/crime/researcher-s-status-questioned-in-jaguar-case/article_e8b9cbd0-602a-5207-9fc2-3e3a06f56540.html

Eisenberg, Christina. 2014. The carnivore way: coexisting with and conserving North America's predators. 1st Edition. Island Press, Washington, DC.

Emmons, L. H. 1987. Comparative feeding ecology of felids in a Neotropical rainforest. Behav. Ecol. Sociobiol., 20:271-283.

Friederici, P. 1998. Return of the jaguar. Biology Digest 36(4): 48-51.

Glen, A. and C. Dickman. 2005. Complex interactions among mammilian carnivores in Australia, and implications for wildlife management. Biological Reviews of the Cambridge Philosophical Society, 80/3: 387-401.

Glenn, Werner. 1996. Eyes of fire. Printing Corner Press, El Paso, TX.

Gonzalez, C. 2002. Do jaguars (*Panthera onca*) depend on large prey? Biological Sciences 62(2): 218-222.

Grzimek, B. 1973. Grzimek's Animal Life Encyclopedia. Van Nostrand Reinhold Company, New York, NY.

Gudger, E. W. 1946. Does the jaguar use his tail as a lure in fishing. Journal of Mammology 27:37-49.

Guggisberg, C. A. W. 1975. Wild cats of the world. Taplinger Publ. Co., New York, NY.

Guynup, Sharon. 2011. The jaguar freeway. Smithsonian 42(6):48-57.

Hornocker, M. G. 1970. An analysis of mountain lion predation upon mule deer and elk in the Idaho Primitive Area. Wildlife Monographs 21:1-39.

Labrona, M., R. Jorge and D. Sana. 2005. Ticks (acari: lxodida) on wild carnivores in Brazil. Experimental and Applied Acarology 36(1): 151-165.

Lemke, T. 0. 1981. Wildlife management in Colombia: the first ten years. Wildlife Society Bulletin 9:28-36.

Leopold, A. 1959. Wildlife of Mexico. University of California Press, Berkeley, CA.

Loomis, Brandon. 2016. Caught on camera: 'El Jefe,' only known U.S. jaguar, filmed in Arizona mountains. http://www.azcentral.com/story/news/local/arizona/2016/02/03/

first-video-el-jefe-only-known-us-jaguar-roaming-mountains-arizona/79775680/

Macdonald, D. 2006. Encyclopedia of Mammals. Oxford University Press, Oxford, England.

Nowak, R. 1999. Walker's Mammals of the World. The Johns Hopkins University Press, Baltimore, Maryland.

Paradiso, J. L. 1972. Status report on cats (Felidae) of the world,1971. Special Scientific Report. U.S. Fish & Wildlife Service 157:1-43.

Perry, R. 1970. The World of the Jaguar. David and Charles, Ltd., Newton Abbot, England.

Rabinowtz, A. R.1986. Jaguar. Struggle and triumph in the jungles of Belize. Arbor House, New York, NY.

Rabinowtz, A. R. and B. G. Nottingham. 1986. Ecology and behaviour of the jaguar *(Panthera onca)* in Belize, Central America. Journal of Zoology, London 210:149-159.

Rudner, Ruth. 2014. Spotted Southern Arizona. Arizona Highways 90(4):48-49.

Sanderson, E., K. Reford, C. Chetkiewicz, R. Medellin and A. Rabinowitz. 2002. Planning to save a species: the jaguar as a model. Conservation Biology 16(1): 58-72.

Schaller, G. B. and P. G. Crawshaw, Jr. 1980. Movement patterns of jaguar. Biotropica 12:161-168.

Seymour, K. 1989. *Panthera onca*. American Society of Mammologists, Mammalian Species No. 340: 1-9.

Wagner, Dennis. 2012. "The cat, the captors and the cover-up." Arizona Republic Series: "Macho B: Last Roar of the Jaguar ... *www.azpbs.org/arizonahorizon/play.php?vidId=5287*

Wagner, Dennis. 2012. "Cover-up amid Celebrations." Arizona Republic Series: "Macho B: Last Roar of the Jaguar ... *www.azpbs.org/arizonahorizon/play.php?vidId=5287*

Wagner, Dennis. 2012. "New UA Study Team Has Ties to Macho B's Death." Arizona Republic Series: "Macho B: Last Roar of the Jaguar ... *www.azpbs.org/arizonahorizon/play.php?vidId=5287*

White, Mel. 2009. Path of the Jaguar. National Geographic 215(3):122-133.

http://www.azgfd.gov/w_c/jaguar/MachoB.shtml.

https://en.wikipedia.org/wiki/Sierra_Madre_Occidental

"Skyislandschiricahuamountains" by BAlvarius - Own work.

https://en.wikipedia.org/wiki/Dos_Cabezas_Mountains

http://www.rainforest-alliance.org/kids/species-profiles/jaguar

http://www.sanpedrosun.com/crime/2013/12/21/two-jaguar-hides-confiscated-belize-forestry-department-san-pedro/

http://www.catsg.org/catsgportal/cat-website/20_cat-website/home/index_en.html

http://channel.nationalgeographic.com/wild/big-cat-week/videos/jaguar-tracks/

http://dx.doi.org/10.2305/IUCN.UK.2008.RLTS.T15953A5327466.en

http://www.tucson.com (AZ Daily Star)

https://www.northernjaguarproject.org/who-we-are/

Illustration Credits

https://www.flickr.com/photos/45131411@N06/15702710631/

http://www.livescience.com/27301-jaguars.html

https://en.wikipedia.org/wiki/Dos_Cabezas_Mountains

www.hcn.org

www.iucnredlist.org (Present)www.jaguar.org.br (Former), CC BY 3.0, https://commons.wikimedia.org/w/index.php?curid=10389837

Jaguar Swimming Photo by lukedakin on Flickr

http://www.alamy.com

http://www.ecoexpeditions.no/america/brazil/jaguar-kill/

http://www.arkive.org/jaguar/panthera-onca/image-G15777.html

http://www.FWS.gov

www.amazon.com

www.wikipedia.com

www.images.google.com

https://www.northernjaguarproject.org/who-we-are/

About the Author

Mary Jo Nickum is a retired science librarian. She has a B. A. in English education, a master's in librarianship and an interdisciplinary master's in fisheries science, anthropology and education. Her first book for children was a chapterbook, Mom's Story, a Child Learns about MS. She specializes in writing science for children.

Mary Jo lives in Fountain Hills, Arizona with her husband. She has two grown sons.

Made in the USA
Monee, IL
05 November 2026

67384983R00064